SPRING FOREST

Geoffrey Lehmann was born in 1940. He graduated from the University of Sydney with a degree in Arts and Law. After working as a solicitor and university lecturer, he is now a partner in an international accounting firm. His first book, *The Ilex Tree*, was written in collaboration with Les A. Murray, and subsequent publications include volumes of poetry, an anthology of Australian verse (with fellow-editor Robert Gray), a novel (*A Spring Day in Autumn*), a book about Australian primitive painters and a standard work on taxation law (with Cynthia Coleman). He is married, has five children and lives in New South Wales.

SPRING FOREST

Geoffrey Lehmann

ff
faber and faber
LONDON · BOSTON

First published in 1994
by Faber and Faber Limited
3 Queen Square London WC1N 3AU

Photoset in Sabon by Wilmaset Ltd, Wirral
Printed in England by Clays Ltd, St Ives plc

All rights reserved
© Geoffrey Lehmann, 1994

Geoffrey Lehmann is hereby identified as the author of this
work in accordance with Section 77 of the Copyright, Designs
and patents Act 1988

*This book is sold subject to the condition that it shall not,
by way of trade or otherwise, be lent, resold, hired out or
otherwise circulated without the publisher's prior consent in
any form of binding or cover other than that in which it is
published and without a similar condition including this
condition being imposed on the subsequent purchaser*

A CIP record for this book is available from the British Library

ISBN 0–571–17246–6

Contents

ACKNOWLEDGEMENTS ix

PART I
The old house

Getting started	3
Photographs	4
Hunger and fear	6
Ex AIF (Australian Imperial Force)	7
The old rifle	9
Tools	11
Poverty Ridge	13
Noxious weeds	15
Down at Hickey's	17
Supper with a black snake	19
Sparrows	22
Jack	23
Chrysanthemums	26
The future of the past	28
Uncle Pat	29
Tommie	32
Mother Church	34
'Menindee'	36
Bird-watching with Mr Long	40
Hens in the saltbush	42
The light on the ridge	44
Kangy Angy	46

Lines	48
Weather report	51
The Spot	53
Shifting gate posts	55
Beginnings	56
Against incendiaries	57
The pressure lamp	59
Life chains	61
Driving at night	63
Kurrajongs	65
Questions for my horse	66
Outdoors at night	67
While fetching wood	68
Postcard for the National Rifle Association	69
My daughter	71
Music	73
Calves	74
The amateur astronomer	76
The things	78
The thrift of tulips	79
Questions for a winter night	81
The old bath	83
Impromptu	84
The Evening Star	85
Witnesses	86
Matt Manion	88
Heat	90
Flowers in the heat	92
A letter from the place of pines	93
Water from my face	95
Iron and calcium	96

Alpine herbfields	98
The meat safe	100
Baking at night	101
With the stars as my bed-lamp	103
The Palace Hotel	104
The small bat	106
Paperbarks	107
In praise of fruit	108
Their day in town	110
The Spring Forest	112
Planting trees in old age	113
JMJ	114
The happy hour	115
Jack Thompson	116

PART II
The new house

Night thoughts without a nightingale	123
The new house	126
Electricity	127
Man and animal	128
Someone	130
The O'Brien brothers	131
Harry Adams	132
Bush kitchens	133
Killing	135
Drying out	136
Jim Long and John Manion	137
The Lachlan	138
Ancient theft	144

Old Testament country	145
Gladstone Watts and the crystal set	146
The two wethers	148
Poverty bush	149
Law	150
The golden wall	153
Prickly moses	156
Gilbert and Hall	157
Jack again	161
Not yet found	162
Invicta	163
The two tents	164
Botany lessons from Mr Long	165
The end of the Pacific War	166
A party of star gazers	168
The Daisy Picker	170

Acknowledgements

This book is spoken through the voice of a living person, Ross McInerney of Spring Forest, Koorawatha, and much of it is based around his life and stories.

I am grateful to my wife, Gail Pearson, for her support and her suggestions. Christopher Koch and Robert Gray read the manuscript in its final stages and gave invaluable help in revising or removing weaknesses. The poem 'Gilbert and Hall' is based on material from Edgar F. Penzig's 'The Sandy Creek bushranger: a definitive history of Ben Hall, his gang and associates', 1985, Lane Cove, NSW, Historic Australia Book Publishing. Some details of the poem 'Prickly moses' are from an article by Janet Hawley.

THE OLD HOUSE

Getting started

When we first came our house
was two weatherboard rooms
in a bare paddock.

I was just back from a war.
There were no trees
and I chose the name 'Spring Forest'.

It was dark when we drove up
and lit our pressure lamps and unpacked.
Our children found potatoes sprouting
on the wire mattress of a large iron bed.
What were they doing there?
my daughter kept asking.

We burned ironbark
in the old brick fireplace,
rubbing etherized hands into warmth.

At dawn Sally and Peter were out
calling in the frost, exploring.
A long icicle hung from the tank.
That day five cars passed on the road,
and the children ran out every time.

Photographs

My wife's the daughter of a professor
and married the winner
of the brick holding competition
at Cucumgilliga School picnic.
Sally, aged twelve, photographed us men
in various postures of expiry
tilted over backwards, or bent forward
holding a brick in each hand in the heat.
But there I am bolt upright, holding
my bricks out to the sun.

My wife born in a large house has come
to our two rooms of aged weatherboard
and an added one of iron.
You can see cracks of sky through the ceiling –
and there's no power,
so our evening conversations take place
by the greenish-yellow light of pressure lamps –
and there's no water laid on,
so my wife washes up in a plastic dish.

Olive, do you regret your life
of photographing country weddings,
and recording local children in silver bromide?
Do you regret the calf born at three in the morning
when I tramp out into the frost?
Or our two children

playing with rocks and broken glass
out in the hot scrublands,
running and calling
to imaginary playmates?
Their garden of jonquils
withered in the first week of summer.

Olive at Christmas wrapping gifts,
inserting notes:
in a book, 'Ross with love from Snotty' (the cow),
'Sally with love from Marco Polo
and the guinea pigs' on a box of chocolates.
A pocket knife for Peter given by the geese.

Hunger and fear

My laboratory
is the dust where I stand,
the sulphur smells of the farmyard.

Your tests show fear
is stronger than hunger.
Maybe true of a laboratory animal,
bred so he's easy to handle.
But try the same trick with farm pigs –
too big and difficult for white-coated technicians.

When their own grass is shrinking,
and the next door paddock is green,
pigs will gather
away from the electric fence, and scream –
in their minds they are already burning.
Then they charge.
Small ones slip under, and big ones,
tangled in wire,
wriggle through – screaming as it crackles.

We are like farm pigs, half feral,
and the fences can't cope
with our numbers.

Ex AIF (Australian Imperial Force)

An invitation to poverty –
soldier settlement farms.
We were back from a war.
Our government said: the young men
saved the country,
they shall farm the country.
There was a stone office building –
I don't know which city –
and a fan was revolving
from a high ceiling
and someone drew a line on a map.
This line became my five hundred acres
and a standard form letter
inviting me to poverty.

No one added up how many acres
make a living,
and I mortgaged my life
to a low interest loan
with a thirty-year term.
But my eyes were open. I signed.

The large properties, broken up
mourn their lost parts.
They will never be whole again.
Small fibro houses
have grown on the fragments

like mould on bread
and dream tenaciously
under the stars of the antarctic night.

We do not starve. We trade favours.
I go away to earn.
You water my animals.
You go away. I water yours.
A countryside of absenteee peasants.
But we return to our mortgaged acres,
our lives of scratching to pay bills,
my life I cannot reject
of squatting on this veranda
as a rainbow lorikeet pecks my finger
taking bread from my hand.

The old rifle

In the long school holidays in summer
I'd be out in the orchard
with an old rifle Mr Long fixed up,
shooting at rosellas
that were raiding fruit.
As each bird fell I'd watch
where the blue and red flickered down,
then I'd drop the rifle and run.
That way I stocked my aviary
with broken-winged rosellas.
And somewhere in my childhood
I dropped and forgot that rifle.

A year of grass grew over it.
Men were working in the orchard one day,
and my brother, the dentist, four years old,
was playing in the grass and found the rifle,
rusted all over – a wreck –
as though it had lain there for years.
My brother knew how to hold a gun
and pointing it at Jim Long, said,
'I'll shoot you Mr Long.'
He said, 'Oh don't shoot me, Barry –
shoot Bill over there.'
Barry pointed the gun at Bill.
'I'll shoot you Uncle Bill.'
'Don't shoot me, Barry,' Bill said,

'Shoot Ted here.' And Ted said,
'Why not shoot Jip?'

Jip was a good sort of dog,
my black and white fox terrier cross,
who was racing around the orchard,
looking for rabbits.
Barry dropped to one knee and squinting took aim.
Jip dropped dead on the spot.
They buried him, telling no one,
but in their haste
made the hole too shallow,
and a few days later the story came out
when the fowls scratched him up.

'You know, Barry's quite a fair shot,'
Mr Long said,
out in the bush with Barry and me.
'My word I am,' said Barry.
'I can hit anything.'
'Can you, Barry, well – see what you can do.'
Barry took the rifle,
went down on one knee
and aimed at Mr Long's billy hanging
from a distant branch.
He fired,
and a stream of brown tea came spurting out.

Tools

Man's tools
are the last stronghold
of something ancient.
You can fool the consumer
but not the workman.
He'll make use
of a newer, more powerful tool,
but the brace-and-bit doesn't change.

In department stores
with their wilderness
of veneers, synthetic wood and plastic brass
give me a counter
of hammers with real wood handles
or spirit levels of solid wood and brass –
or my cattle cane,
its handle plaited with hide
hanging on the veranda
among my rifles.

I spray my tree
with a long thin pump of brass
that can reach among the branches,
elegantly,
a design that's not changed for years.

On a hot day
the metal chills your hand
as the spray flows through.

The tools are tenacious.
My spanners will be able to take
the nuts off spacecraft.

Poverty Ridge

'Poverty Ridge,' Mr Long labelled it,
'the loneliest camp in the district.'
A time of drought and bailiffs,
a red MG with a loud muffler
patrolling the hills,
my brothers hiding under portable bushes
in a paddock after dark,
a lavatory seat banged down as a signal –
a time so dusty
my wife took to brushing the dogs with a broom.

Sally at the windy end of winter
would rub the wattle buds
(just on the brink of blooming)
so the yellow whiskers popped out
while we went looking for lost sheep
in the dry, cold paddocks.

My sister rabbit trapping.
Her children carry the corpses tenderly,
except Kellie, who bangs them on trees and rocks,
holding an ear in each hand,
twirling the corpse over on itself,
like a grocer with a paper bag.

Rabbit kittens were spared
and taken home to a box,
but the house stank of 'currants'
when the rabbits broke out.

Noxious weeds

Driving through our district in late spring
with my daughter and new son-in-law —
it's not country I would go to look at,
mainly ironbark and box country
broken up with outcrops of granite.
On a bank purple with Paterson's curse
there's an antique sulky
for the tourist who pays a dollar
to see a cow standing in a paddock.

I recite the deaths and accidents
this road has known.
Past Morongla Post Office
I miss that red rambler with the white eye
wiped out by the Shire
in mistake for a blackberry.
'What sort of poison are you using?' I asked.
'They oughtn't to issue you with that.'
The man from the Shire who was spraying
screwed his face up in alarm —
so we still have notified noxious weeds,
our roadside briars
whose leaves in spring smell of apples.

The house is among those trees.
When my son-in-law first came
my brother the dentist with toothbrush moustache

asked,   How does he find the toilet facilities?
meaning either down the creek
or over the hill.
Now my brother asks, 'Does he call you Dad?'

A marriage on my wife's birthday –
we're all born and married on the same day.

Down at Hickey's

You can't hear it in the house,
the wind in the upper air.
But out in the paddock
with just the sky-travelling moon
and your shadow on white grass
it sounds like a distant express train,
gusts of huge pressure,
while on the ground
the ears of phalaris are barely stirring.

My bed has a cover
of red calico striped pink.
On these cold autumn nights
grass is turning into milk,
and on the veranda where I sleep
ironbark seedlings in metal tubes
are pushing out slender blue leaves,
while down among the poplars
on a rusting mattress
the ghost of Mr Hickey sleeps.

My postal address is 'Spring Forest',
but we still talk of
'down at Hickey's'.
I wish I had the teams
of school pupils he used
to weed his vegetable garden

sixty years ago.
Slave labour for country schoolmasters.

The pressure lamp hisses in the kitchen,
and hot cocoa faintly steams.
A green lacewing lands on the board table.
Several times a night
Joe wakes me with her nose
to make sure I'm not dead.

The leaves of my poplars drop
into Mr Hickey's open mouth.

Supper with a black snake

We're sprawled in lounge chairs, lolling on the carpet.
After the talk of weather and crops and blight,
with tea and cupcakes
the snake yarns arrive late at night.

'About to pick up some tarpaper in a shed
I saw a black tail nesting there.
"Jack, there's a snake!"
I said, and went to get my gun.
Jack gathered up the tarpaper,
but not for long.
"You bastard," he cursed, as the shot rang out,
"I thought you were joking." '

'Crouched under the tipped-up cabin of a truck,
unable to move, with a spanner around a nut,
a black snake glided down my shoulder
and into the sump just as my son arrived to see
the tail disappear.
"Why'd you let him do that, Dad?"
"He wasn't botherin' me
and I wasn't botherin' him," I replied.'

'Baling hay, you need a man at the start of the chute
to watch if a snake's coming through.
Collecting and stacking the bale, you find
the snake is packed in neatly,

or just a tail's hanging out. That's fine.
But if it's the head,
as the bale comes down the chute
a friendly yell of "snake on the way" does no harm.'

'Talk about eels, fried snake's not bad.
Chop it up and tip the bits in the fry pan.
They wriggle as though they're alive.
The men all grouch, but it doesn't stop them eating.
It's tasty, fried snake. But for a good feed,
you need a monster that's forty foot long.'

'He was carrying a roll of lino.
It seemed strangely heavy.
Then the cause of the extra weight poked its head out
to see what was happening.
Bill and the snake didn't waste much time
looking at each other.'

'Tigers rear up and strike back
if you attack from behind.
They're slower moving forward.
Let them come at you. Shoot from in front.'

After an hour of snake yarns, it's time to drive
over corrugated roads to a house
where women fret that the doors aren't flush
and veranda boards have gaps.

The venomous snake in the paddock is tolerated,
it is snakes in the home yard that are loathed.

My rifle is propped by the front door.
Its wood is oiled and polished,
and wasps can't nest in the barrel
with its plug of blue tissue paper.

Sparrows

I don't mind if silver-eyes eat my grapes
but sparrows –
aggressive foreigners –
are the plague of my life.
I built a bird bath
and sparrows came,
the miniature thunder of a hundred wings,
and drove the natives away.
But I'm forcing them out
of my silver poplar grove,
poking their scruffy nests down from the branches,
leaving the neat circular nests
of blue wrens and other natives,
setting traps for them on the ground,
a war of attrition on sparrows.

I've no time for birds
with no limit to their breeding
and slovenly nests – like some people.

Jack

I've given up drink for good,
by natural evolution.
Alcohol is for the young,
out of love, in love,
young men chasing the same woman,
urinating by night among camellia bushes,
ramming trees with their cars.
Somewhere there is music playing,
glasses are breaking;
they cover themselves with grass seeds
and mud, teased by girls.

Now I've enough mental furniture
to shift around in my mind to keep it busy,
something the soft minds of the young
cannot understand
who see middle age as a shrivelling,
not a storing away of energy.
So drink is strangely irrelevant;
I say 'strangely', because unplanned.

My house is dry except for the grape-vine
that loops around the veranda.
Jack would have said what a dullard I've become.
'Come on Ross,' – his ghost standing
between the grapevine and me –

'What about a booze up? Olive
and the children don't want you to martyr yourself.'

Jack and his friend Higgins,
as medical students,
swaying on the doorstep of a hostess
before a dinner invitation,
and Jack persuading Higgins, full of beer,
that a pot plant would soak up the lot
without trace.
The pot plant stood in a saucer
that was filling with tepid water
as Jack rang the bell.

Jack crash-landed in the sea
off Vanimo,
helped free his two friends from the cockpit
as the plane submerged,
and drowned himself.

Jack became a small part of history,
a pioneer doctor
bringing civilisation
with a stethoscope and syringe
to the dark people who mourned him.
I was Jack's audience,
the younger farming brother.
Jack,
fatal for girls, fatal for himself,
my mentor and guardian in the city
of wartime parties, floating populations.

The roof garden of Packie's Club
with potted palms, outlines of office blocks –
this is *my city*, a wartime city,
railway stations where we sang
and people smiled at the drunken soldiers;
but now it's gone, the T & G building,
Packie's in the starlight,
my city only exists in my mind.

But Jack bitter, unreconciled,
chasing a glimmer of phosphorus on the horizon,
stands between my grapevine and me:
the hundred possibilities as I stepped
from the train at Central Station.

Chrysanthemums

This is something
about late starters – chrysanthemums.
I suppose we humans
start slowly. The baby
playing with his food, dropping it
even when hungry
is no match for a twelve-month squirrel.

Shortening of days sets off a time switch
for chrysanthemum buds to form.
Dull, intricate weeds,
through spring and summer
they've nothing to say.
But in autumn they burst
in motionless fireworks around the homesteads,
yellow and vinous red and russet,
sagging by wire fences,
dusty by galvanized sheds.
They bivouac in the long dry grass
beside the spacious verandas
as lucerne is baled in paddocks
and basket willows turn yellow.
Unlike the irises and jockey's caps
of summer, living for hours,
or hibiscuses quickly crumpling,
the chrysanthemums select a time
when the sun will not exhaust them

to speak out acrid and dry.
Their blaze persists,
saved up, parsimonious, calculated,
like the last burst of a marathon runner,
as the sun shrinks
giving back all the light they've known,
dazzling as they totter.

What a way to die!

I'm not saying chrysanthemums
are best.
Perhaps I am fonder of the flowers
of spring – the soft scents,
expanding with the warmth.
But I like the way chrysanthemums
answer the cold
with remembered light.

The future of the past

Oddments.

Once my brother Jack crash-landed in New Guinea,
flying to a patient,
(not the time he died);
a plane flew over
and they dropped a carefully wrapped
bottle of rum 'for medicinal purposes'.

Talking in a tent during the war
I heard a stranger in the dark
shout from another tent,
'Hello! Jack, you old bastard!'
(Jack and I had the same voice.)

Lately my daughter woke at dusk
and heard the radio saying,
'A plane has crash-landed in the sea
off Vanimo and the pilot has drowned.'

Not Jack, but his double fifteen years later.

Ageing I speak with the ageing voice
of Jack that was never heard.

Uncle Pat

What despair or contrariness
persuaded Auntie Bridge
to marry Pat?
Old Pat driving his car –
'She houlds the road well,' he would say,
letting go the wheel to prove it.
'Ouh!' he would grab the wheel again in a panic,
yet the steering hadn't shifted an inch.

Auntie Bridge and Pat –
I felt like chasing the old blighter
and ducking his head
in one of his cowshed buckets
when he gave my children some holy medals
and said, ''Ere, take these.
They'll keep people orff yer.'

Pat believed in Adam and Eve,
the snake and the rib-bone, that made sense.
He had no objection to the chariot of fire
and the dead rising from their graves,
that was something he was looking forward to himself,
Pat rising out of red Cowra earth
from a patch of skeleton weed
and not letting Jesus' hand go –
No, old Pat would be taking no risks
in his riding boots, 'houlding' on grimly.

There was only one Bible story
he couldn't quite fathom.
How did Noah
fit all them animals into that ark?
It was hard enough getting cows
into a milking shed or sheep
across a road. How did he round up
an elephant couple,
or even worse, all the moths and midges,
snakes and ants?
Did they have termite mounds on the ark?
How did they stop all the animals
biting and poisoning each other?
Pat's mind ran around in circles,
worried that this one gap in his faith
might keep him cooking in Purgatory,
scratching his head for a million years.

A Pat in 2050 won't be so incredulous
about rounding up
all the species that are left.

We're herding them into our arks
of concrete and chain wire,
the last of our friends, the beasts
who travelled with us through time –
herding them into our zoos and game reserves
before they're engulfed by the Flood –
and the Flood is us.

The man from the Shire
talked me into laying baits,
carrots dipped in tasteless liquid.
And we go a few rabbits –
and half a dozen wallabies with them,
prone in the dust among small native pines.
Next time I'll dig the rabbits out.

The condor going and half
of the animal world with it –

Each species we destroy
is a plane lost off Vanimo.

Tommie

A girlfriend's Mercedes
is parked in the shade of a Cootamundra wattle.
Inside, Tommie (my mother) is doling
out gossip and tea,
quietly proud of her position on the ridge,
looking out over miles of country
but hating the six gates
between house and road.
The cars are gathered for a bridge party,
a thousand feet above sea-level,
among the lichen of high places,
encrusted rocks and rotted fence posts.

Tommie (the ladylike spelling was Mother's)
née Miss Thompson,
married into a family of police-hating Irish,
what did your father do?
Don't mention it to those cars parked on the ridge,
as a fly settles on a tinted windscreen.
My father's nickname for her,
'Criminal Investigation Bureau' –
too close for comfort.

Oh Tommie, the Bush lie detector laying traps,
even for granddaughters.
'Sally, don't you think Barry's legs
are awfully thick?' No answer.

'Now Ross,' she'd say later in Barry's hearing,
'Sally says Barry's got legs
like tree trunks.'

Tommie being driven down
Mount Tomah, eyes shut, huddled forward,
face in her hands.
'Oh Godfather, Barry, we'll all be killed.'

'Dan,' she once said to my father,
'There's two dead mice under your bed,
one in each drawer.'

Tommie,
privately licking the cream from the egg beater,
then emerging to join her ladies at bridge –
but now they're going, it's dusk,
the cars winding down the hill,
stopping and starting at each of the six gates,
red tail lights winking
through mist rising from stagnant ground.
And there are sounds from the kitchen of 'smashing up'
as Father calls it,
who arrives with the tea-towel of peace.

Mother Church

I came into this century a Catholic
and shall leave it with no belief,
like a hundred million others.
Faith like mineral salts
is leaching out of the soil.
A Pope was afraid to speak out
against blackshirts and murderers
and did nothing
as the world drifted into war.
A later Pope, lost in his own dogma,
was too craven to speak the word
that would release the unconceived
from a birth without future.

Transubstantiation, free will –
mouthfuls of nothing.
Mother Church
when our earth cried out
you had nothing to say.

Where is your 'life after death'?
What about the life now?
'Infallible' – 1870.

How silly it all sounds to our ears.
I can only laugh and cry.

Yet if outsiders attack you
I'll strike them down.

'Menindee'

At thirty-three my brother-in-law
is dying
of a hatband that chafed a mole.
With seven children –
three his own,
four others the children of a dead man –
he won't give in.
Married at last to the woman he battled for,
still begetting children,
after the brandmark showed,
he insists that unwilling surgeons
cut as each secondary appears;
excoriated and scarred
in a one-man cosmic war
against death.
In the last year he has built a dam,
purchased a capacious deep freeze,
set the fences in order
and the women are saying their prayers.
My sister is in a state of collapse.
This year with spring
cruelly lush and wet
his false lucerne tree flowered profusely.
The lilac he planted,
and other costly trees and shrubs
are just coming on,
but my sister, the horse rider,

is a ghost of her beauty,
lined and haggard with his war.
The children she knitted and sewed for
are growing unkempt like weeds,
and the house
the cream wooden cottage on the flat
tucked away in an elbow of creek
among pepper trees and haystacks
is an empty shell hole
where they subsist
in the lushest summer for years.

There are spiders on the fibro veranda
and cats germinating under the boards.
Clothes are scattered on the laundry floor.
Two summers ago newly married
they would lie there, idly joking,
soaking up coolness from the concrete slab,
gazing at the ceiling, dim green
from the garden's glare.

This is chronic country,
terminal perhaps for some.
After ten years of drought and waiting,
the lush seasons have come
and wool has collapsed –

Terminal for the big establishments
with big mortgages
building up their flocks for years

and borrowing, and now
they're selling prize rams for carcases.

We are sheep farmers
and nobody wants us –
driving at night through paddocks
the eyes of sheep in our headlights
are green incandescent jelly,
shaking and moving away as our tyres bite through gravel.
Men reject the soft wool
which comforted us in the Ark,
preferring the loveless synthetics,
false economics.
But beware, your children
will curse you for letting dodoes die,
half of the natural world perish,
and when our flocks have dwindled
you may not entice us back.

This planet which tries to house
half of the men who have ever lived
wants no one in particular.
It does not want you, either –

We are all sheep farmers.

Myself I wage no cosmic wars,
I travel light
with my five hundred acres,
half of it uncleared, kangaroo country
because no one wants

what it would grow.
With my bees and yellow jonquils
and journeys with a carload of calves,
trading in a small way,
I survive.

My sister still desperately beautiful
rides the boundaries of their big establishment.
One of her children sleeps with his eyes open.

Bird-watching with Mr Long

'What's that bird, Mr Long?'
'That's a chipper.'
'What's that small bird over there?'
'That's a fly-bird.'

There's a forest I'll never see again
where birds with exotic names
whistle to each other,
flashing blue and scarlet
as they dart and fan their wings.

'What will you have for breakfast, Mr Long?'
asked my father.
'I could eat the leg of the Holy Ghost,'
replied Mr Long (meaning toast).
'I would *not* have expected that of *you*!'
said my father with ice.
But Mr Long was rarely put out.

On a wooden chair by my bed
there's hot cocoa I'll drain fast
because these autumn nights
are taking the warmth out of things
as they loosen the poplars' yellow leaves.

Then I'm going on that journey
Mr Long always promised

through the spinifex
with a covered wagon and cockatoo,
cooking fish on river stones,
to Palm Valley and its wild blacks.
'What's that bird, Mr Long?'
'That's a parson bird with the white collar.'
'And that one over there?'
'That's a grey hopper.'

Walking all day
out on the western plains, Mr Long
could sustain himself
with a line of trees on the horizon.

Hens in the saltbush

If he was a man of spirit
you ate your enemy after fighting him.

But the flesh we eat now is insipid,
blind, force-fed calves,
and mass-produced hens
with their egg yolks the colour of custard.

My egg yolks are rich yellow,
so deeply yellow they're almost bloody
and city people get squeamish.

Mind you, my hens eat everything!
Dirt and dung and wattle seed.
They roost in the shells of rusting trucks
or on a low branch of saltbush.
(I introduced it – Spring Forest's not desert country yet.)
And my dogs sometimes eat hens
and so do foxes.

That's bad for the profit and loss accounts.

I've been a great hater in my time –
battery-raised hens and plastic wood,
cities destroying the landscape –
but now I rein my hatreds in.
The crusaders

wear themselves out hating,
and what matters – their own minds,
where it all begins and ends –
is forgotten, they're so busy hating.

Plant the Spring Forest,
to start with, in your mind.

The light on the ridge

Throwing stones in the creek forty feet below,
careless in those remote hills,
my brothers and I heard a rock
not thrown by us hit the creekbed.
We threw more, and paused.
Again a ghost rock
knocked on the pebbles.

We peered over the boulder's brink.
There was Father,
sitting on an unreachable ledge,
smoking his pipe.

Medical science preserves our bodies for a ripe old age
in the men's home.
My father, eighty-seven, rides a horse
and tells me of a home for the aged
where they are locked in their rooms at night.

I watch for his light on the ridge
five miles away,
living by himself now Mother's gone.
Peace, heavenly peace, he used to call it,
while Mother was away,
and now the heavenly peace is permanent,
and peach trees have grown to obscure the light.

My father, the grandfather of many,
takes no sides, commits himself carefully
to no one,
white-haired diplomat, peace-maker.

At night investigating moths in his pressure lamp –
white with orange stripes, that deserves a prize!

A master of courteous truisms –
but why did he keep peacocks and a monkey?

Like underground water,
that I struck last week, three hundred feet down,
dissolving in and out of landscapes –
Father.

Kangy Angy

At Kangy Angy on the coast
someone has left a stand of cabbage palms
in the middle of an orange grove.

I drive past and don't stop to admire,
in case I'm spotted
by a man ready to set up a hot dog stand.
Next he telephones for a concrete mixer
and trucks with bricks,
and six months later
I find the Cabbage Palms Motel
and Orange Grove Fast Food,
with litter bins
painted conservationist brown,
and signposts pointing to my stand of palms.
They've left no undergrowth or whiskey grass,
and Kangy Creek is full
of canoes and motor boats and tourists –
harmless voyeurs of leaves and bits of rock.

No, if I stop, it will be late at night,
when there are no cars
and only the odd semi-trailer
slews past, with roaring lights
and a driver half asleep, high off the road.
He will not see my shadow
slip across the wire fence,

inhaling the scent of citrus,
and vanishing among the palms,
that lived
in the rich soil of these flats,
before there were men.

Soon the only places left
with any self-respect
will be the ugly towns, the plain landscapes —

Lines

A vertical line through our roof
would intersect
with stars somewhere in space.
(There are other Spring Forests in the sky
and children crying. The stars
are a million mirrors of the earth.)
Closer to home this line
might bisect the moon's molten core,
and pass through
radiation belts,
the ozone filtering out the ultraviolet,
a tawny frogmouth flying
with a moth in its beak,
frost on our galvanized roof,
a kerosene pressure lamp perched on a book,
various texts on animal husbandry,
some short stories
and a cherrywood pipe
I have lost and not yet found
(that's wishful thinking –
I probably lost it in some paddock
and a tussock has grown over it)
down through pine floorboards,
a ginger and black guinea-pig asleep
beneath the house
and into red Koorawatha earth,

earth with only one need —
water for the green life chains.

If I tired of vertical lines
I could draw a horizontal line
through this fire of ironbark logs
(with its two sounds —
the billowing and beating
of rushing blue-red air,
and the dry cindering and splitting
of timber)
a line extending through the curl
of steam from the iron kettle
warming on the flagstones,
through my moleskin trousers,
as I sit on an old car seat from the Morris
(my favourite low-level armchair)
just missing Olive's legs
busily gathering tea-things,
on through the bedroom with its black piano
carved with flowers and mandolins
(how the steel strings and sounding-board
wince in our draconian ranges of temperature —
the felt hammers decayed
when my wife the musician
married me and a farm)
through the weatherboards
and a stand of red geraniums,
on past the trunk of a giant dead wattle
(I don't remove old friends,
as birds like to perch in bare branches)

through the chicken-wire enclosure
I keep around the house and garden,
past some dogs and a fruiting fig-tree,
past the cough of a fox.
I jump up with a gun and that's where that line ends.
But it's no use. Try shooting ghosts.
I come back inside.

Drinking a cup of cocoa
I draw a circle around the house
starting with the metal windmill
and the creeks where the ducks paddle,
but that's too wide.
I'll start my circle in closer
among some grass. It collects a hen
in a crater of dust, continues
through the bee-boxes with their new white paint,
on past my antique steamroller
'the slumbering giant'
and then I fetch up against that fox again –
or is it my mind?

We have cosmic rays and cow manure,
flowers and a rusting dry-cleaner's van,
but there's no line around here
that will intersect with a decent toilet or bathroom.

Through the dimensions I do not understand
I move
a column of living water.

Weather report

My father's a still day
smoke rising vertically in the calm
from a distant horizon.
There are high cirrus clouds,
mare's tails, thin streaks of ice crystals
combed across the sky.

My sister's light air,
smoke drift, a faint breeze
you can feel on your face,
and leaves rustling under
a deep peach sky at sunset.

Mr Long is one of those small clouds
that sit on top of mountains,
a wry companion,
scud that rushes across the sky in a storm.

My mother is cumulus cloud,
brilliant white and puffy in fine weather,
billowing and changing shape with her mood,
while leaves and small twigs are in constant motion.

She starts raising dust and hen feathers,
the wattles around my house begin to sway,
and telegraph wires are whistling,
as mother becomes a gale.

Smoke venturing from the chimney
is shredded into nothing.
Our old white horse (that's me)
canters around the paddock, wondering
why the sky's become so black and blustery,
as branches are breaking off trees.

Then it's sunny again,
and the glistening hillside
is my one-month-old granddaughter's face.

I've thought of all this,
on a summer night, silent
except for frogs which mean water,
smoking by myself on the veranda.
And the sky is lit by static lightning,
violet flashes. Jack.

The Spot

A child was killed crossing a road.
A neighbour, kind and honest in his way,
says, 'Ross, do you want to come and look at the Spot?
Ellen and the kids are going to have a look.'

When cars crash the traffic slows
to a snail's pace as passing drivers
crane to look at the wreckage.
We are absorbed by scratching our scabs,
or taking a quick look backwards
at our turds.
Making happy noises
babies try to scratch out each other's eyes.

There is part of our nature
where morals and logic are irrelevant,
some marshland of the brain.
In another latitude
Yellow Thunder, an ageing Red Indian,
was beaten up and thrown
with no trousers into a crowded dance hall.
Later he was wandering the streets.
His young tormentors packed him into their boot,
and dumped him at the edge of the town
to die.

In most of us ancient malice
has atrophied into words or thoughts.
Listening to music or holding the hand
of a woman or child you love,
where do these pots of boiling fat belong?
They are not the meaning of our dream
(as some say).
Criticism is useless.
They simply exist.

Shifting gate posts

Even earth is not stable,
and gateposts tilt,
so to keep the gate shut
the chain grows a forelock of wire.

Our faces and bodies
are much more makeshift,
and wire won't always bridge the gap.

One of the pallbearers said,
'By George that box was light.'
Crouching on the veranda edge in riding boots
I press the plug of burning tobacco
with my thumb
and smoke blows away in the night.

I come
as an observer, not mourner.

Beginnings

In the dip of a paddock seven tiny broken eggs
mottled brown and white
(powder blue inside)
announce we have seven more quail –

my daughter
standing on the top branch of a pine tree –
we called her sminthopsis,
marsupial hopping mouse –

myself aged eighteen months and naked
with a silver hairbrush on a jetty
welcoming Uncle Bill from the first war –

these signs
beckon us forward not back –
we drink from vanished springs.

Against incendiaries

Clearing and burning off –
from my father's place the fires
on the plain look tiny and remote,
as dusk closes in.

Driving home that night
as we come round a bend
the fires are on us, in close-up,
a huge stump blazing in the dark,
the felled logs burning
in the bulldozed blackness,
deserted and eerie
as a flying-saucer camp.

Tossed on the hearth at home
the empty skin of a persimmon
glows orange as the fire itself.
Paper, cotton, matter that has lived
burn cleanly, leaving a pale
ash ghost.
Plastic and synthetic cloth
bubble, reduced
to fuming chemicals.

I was going to build a fire in my mind
where I could burn
all the trash of our world,

lurid newspaper headlines,
our constant exposure to violence,
machines designed not to last,
advertisers, sterile cities.
I wanted
charred earth, stubble
for a clean start
under stars scrubbed large and new.
But my firesticks got lost somewhere
by an old pear-tree
or behind a tumble-down shed.

I've decided not to burn
for my beliefs.
Heretic and inquisitor
feed on each other's flames.
They'd incinerate our world too.

On the coast they burn off paddocks on a whim.
Our feed is too scarce:
we only burn scrub we're clearing.

Outside hearths and combustion engines
fire is something I suspect.

The pressure lamp

With Olive away, the house is in darkness.
My feet fume with the cold.
There is nothing, no room, no house, just freezing darkness
as I rummage
for a match.
I am dead. We all died
on the same day
and are buried by the river
our chins tilted upwards
still sprouting beard.
We are all dead under the rotting leaves
under the trees dying of mildew.
I manage to strike a match
and place a ring of blue fire
around the stem of the pressure lamp.
The smell of methylated spirits.
The mantle (silk charred to white ash)
trembles and smokes with orange heat.
I pump. The lamp and the room
hiss into light,
a resurrection of familiar cupboards,
the violin on the wall, books at all angles,
the straw mat in the middle of the floor.
I have made my own light.

I place paper and kindling twigs
on the hearth – and fire,

my second need, has been established.
I make my own climate
(my domed microclimates
on the icy moon) –
one step further than the warm-blooded animals,
two steps further
than the lizards and insects
who are cold or hot as the day.

I move from light to warmth,
to food and sleep,
the last of my simple needs
of a winter night,
and the house is in darkness, silence,
waiting for the chains of actions
to start again.

Life chains

Life chains –
a lamp of incandescent silk,
so much light from a small silk net
(see yesterday for an explanation).
Life chains –
sheep farmer, wool scourer and retailer,
the two spirals of the DNA molecule
twisting around each other.
Some ancient mind read its message in our dust
without an electron microscope
and saw the caduceus,
the two snakes writhing around the wand
of commerce.
Life chains –
so tricky to tamper with,
no beginning or end,
stretching beyond the brief flash of our life.
Outside our room of light
are genes we do not understand,
systems too fragile to observe.
Chains in more ways than one,
binding us, hard to break –
the revolutions kill
and do not make us free.
But if the life chains lead to death –
the growth economy
exploding on itself,

half of mankind starving
in darkness with no match and no lamp –
we must break
and rearrange
the life chains
by intelligence, will, perseverance
before they have gutted our planet.

On top of my refrigerator there are some eggs
and a pair of earphones.

Life, the silk mantle that started in a caterpillar's stomach,
incandescent –
(I was going to say burning; but that's not accurate.
It burns only the first time it's lit – for a second.)
Jarred, it disintegrates.

Driving at night

Driving through thick bush
alone – mist scatters in my headlights.
The death of a parent.
The earth loses its heat
by long wave radiation at night.
When the sky is clear
the long waves go out into space –
sweltering Christmas dinners with my mother
eating her family with pudding and brandy,
stentorian gossip,
the panic in the bushfire –
are leaving the earth
and shall not come back.

The earth loses its childhoods,
wood houses with their hearths and willows
flow away into the sky,
fathers and their horses,
mothers with iron pots
are going, and wives
who were warm
when dew formed on tin roofs
leave a crater of coldness in their beds.

There are no clouds to stop them.
The long waves leave us
feeeling nothing.

Movement of air in the hills
turns dew into mist.
On the plains it's dead still
getting colder and colder.
A frost for my mother.

Kurrajongs

Kurrajongs,
scarce in any landscape,
one among a hundred trees,
with your salad green cloud of leaves,
and your barrel trunks,
water-storing, bark of elephant hide.
In drought when other feed
has shrivelled,
we lop your branches
as fodder for starving beasts.
My father visits and tends each tree,
a staff of life,
slow growing, large only at a great age.
The gums and wattles are hybrids
that vary from shire to shire.
The kurrajongs do not change.
They carry their standard gospel
across a hundred thousand square miles,
temperate in our fierce landscape,
on drab plains
these shimmering bubbles of green.

Questions for my horse

Music is unevennesses
of pressure on the ear-drum.
Sight is the vibration
of rods in the eye.
My dog's called Joe.
Meaning to ask for Ock her son
I asked Mrs Wearne
'Where's Olly?' (her dead husband).
'You tell me,' she said.
Waking in winter –
a big bush cat was sitting in the starlight
scratching at green parrots in a cardboard box.
And where was Olly?
You tell me.

Outdoors at night

It's surprising
that the universe is able
to look at itself, from end to end,
the near fires and remote fires
burning in a clear vacuum.
I stand under the grey antlered limbs
of my dead box tree at night
and watch the stars
signalling to each other.

As light a million years old glimmers on me
I ponder, among my white bee boxes,
how much more likely
that space should be opaque
an obscurantist's delight,
a vast sponge to be lost in.

No . . . the stars announce
their presence over huge distances
to my yellow tractor, a beetle and myself.

While fetching wood

Galaxies receding –
their light shifts to red –
The horn of a train approaching blasts us –
when it's passed, the pitch drops.

My neighbour sent his horse to the knackery
when its working life was over.

The music or person we cease to love
suffers a red shift, a drop in pitch,
but does not change –
only our position.
They go on in time, the objects, the persons,
with no need for our say so.
How cheap to think of their worth
in terms of ourselves – trees,
earthworms, old sheds and a snake
lacing through a stack of logs
I can see into, black against light.
And I don't kill it.

'Good day, old horse,'
I say to my neighbour
on his way to the knackery.

Postcard for the National Rifle Association

They're out shooting again tonight.
Driving past they shine their spotlight
through my naked silver poplars
on to me in bed on the open veranda.
(For a second I'm on stage!)
They're charitable.
Last time I found a dead mangy fox in my mailbox.

Guns are for cowards –
too scared to grapple with their enemy
face to face.
The flabby finger squeezes a trigger,
the eyes look away
as the quarry falls.

There's no mystique about killing sheep,
no sheep slaughtering clubs,
but Hunt Club secretaries tell us
we are squeamish about killing
the meat we eat.

Well, you don't find men in red coats
killing meat for their sausages,
or the Duchess of Buckingham
slaughtering hens
at the chicken processing factory,
while her lady-in-waiting hoses away the blood.

I stopped some fellow shooting
on my property,
took his two hundred dollar rifle
(a very fancy model)
and with bare hands broke it across my knee.

Traveller, do you shoot?

My daughter

Sally's poem.
'My father seemed so strong
and well able to survive.
But I waited in an agony for my mother
(afraid she might die or get lost)
when she went over the hill.
I climbed up trees to watch for her,
sweating with fear.
At night when father went to bed
I listened to her shoes
scraping the linoleum, her hands
stacking dishes or turning a page.
To hear her every movement
I lay
breathing through my mouth.
(If I breathed through my nose, air
made a rushing sound in my nostrils.)
I kept my head still
so my hair didn't rustle in the pillow.

'One day she had been away so long
from a trip to the bushes
I climbed a silver poplar
and saw some animal larger than a dog,
a brownish-chestnut colour,
in a field of wheat, but no mother.
I couldn't quite see its face.

But this thing had eaten my mother, this was certain.
Then it was gone.
I came down, an orphan,
and found my mother, a ghost
on the veranda talking to Peter.
There was nothing I could do with my mind.'

Music

This house hasn't known much music
except Sally sitting in the dust,
tightening the wire strings of a bee box frame
and plucking them.

At night the trees rush with different sounds
or a bull is restless
and dogs interrupt the darkness,
these are a sort of music –
or sleeping in a travelling car my ear
listens to the change
from bitumen to gravel and back again,
the chassis vibrating.

Young, I needed an occupation,
felt myself going mad without it.
Older,
I find music in anything,
sounds of nature, personal idiom,
doing nothing –

Calves

Some musical intervals survive
from when the ice sheets began retreating.
Feet travel over grass.

I'm travelling with a carload of calves
at night from Bega.
Shivering by the side of the road
my breath scatters over dry grass.
Trading
in the soft bones of new lives,
I come with a carload of hope
and inquisitively sniffing noses,
drink steaming coffee at a service station
under fluorescent light,
and drive on.

In colder regions (I reflect,
passing through a patch of mist)
the animals become scarcer and larger.
My tall frame is on loan
from some disgruntled ghost who lived by a peat bog.

The farmers who will buy my calves are asleep.
Past midnight most of the lights
in the district are out.
In my back seat I carry
next year's herds

(how they'll run to meet me at the fence
and butt me in the waist) –
the latest batch from long blood lines.

I travel through a tunnel of trees
over pale gravel,
with my lights on high beam,
the only moving object for miles.

The amateur astronomer

People no longer believe what poets
or ministers of religion tell them
unless their senses say it's right.

All those cancelled Utopias,
and syllogisms that just didn't work –
relieved of the incubus of trying to believe,
I walk down my father's mountainside
one night in July, unprotected –
nothing between me and the wind
blowing from Antarctica,
nothing between me and the stars
glimmering at the bottom of space
(the Antarctica of the sky).

The wattles and native pines seem to enjoy
this cold wash of air,
this lack of illusion.
(I do not say disillusion.)

In every large city
there are a hundred or so amateur astronomers
picking their way through the sky.
I'm not that sort of fanatic.
For years if I looked through my antique brass telescope
all I could see was a broken lens
and cobwebs.

Now it's restored
I can train it by day on my father's hill
and see someone hanging out washing
up there, five miles away.
Among tussocks and small blue daisies
invisible in the dark,
I can see the moon as large as a plate
and the rings of Saturn.

Dabbling among solar systems,
I'm as happy as a cow in fresh grass
with no knowledge of botany.

It's time to eat, Olive tells me.
Yes, Olive, I'm coming in
with the moon as my dinner plate.

The things

My daughter's Bunnikins plate
has had its garden of rabbits
rubbed and washed off.
When she grew up
we lent it to a shingleback lizard.
Now it's my granddaughter's plate.

The things are hallowed by use,
old purple glass in the dust of my yard,
a breadboard of myall wood
with the clear smell of violets,
and a flower pot from my aunt's dead garden.

Things in the mind
become emblems and logical pathways,
a father with a shotgun and Mr Long hiding,
the green eyes he never saw again
and shoes he left on the roof.

The thrift of tulips

It's the warmth coaxes tulips out
and makes them flower,
yet they languish in our warm climate.
I'm convinced they love cold
only because the great sterilizing sleep
of a northern winter
kills tulip parasites.
The poisonous bulbs, the jonquils,
thrive in our district
(round and fat like spiders' stomachs)
but the harmless tulip bulb
(succulent shaped like a tear) survives
only when the ground has been civilized (for tulips!)
by ice.

We tried tulip raising only once,
a row of green buds
and one day a bright red flower,
but the calyx in which it came
had vanished.
More flowered – again this conjuring trick.
Nothing clung to the stem
or was scattered on the ground.
Then one day we watched
a green calyx blush
and deepen in a day to red.
The calyx was the petals.

What a masterstroke
in the geometry of growth!
But I've planted no more.
I only grow what's happy in our ground.

Questions for a winter night

Why does a cockroach
look like a pressed date?
How does a snake
locomote across water?
You'd expect it to sink like a strip of lead.
You ask the skeleton of a snake,
the fine light rings of bone
among dry grass, they provide some answer.
Why do people shake their heads to say no?
I can answer that one.
Chidren shake their head from side to side
to avoid food they don't want.
The nod is the acceptance of food.
Why did Sally place her foot wrong
so Cucumgilliga Primary School
had to start each folk dance again,
her embarrassment multiplying her errors?
Why didn't the twenty-four feet
of Cucumgilliga Primary School dance in unison
like the segments of a snake
oscillating across water?

Why is the moon
(high above my radiata pine)
a cross section of frozen apple,
and how did it arrive there?
It's a winter night

and ice is forming at the bottom of miles of still air.
My breath hangs in a cloud.
Why am I on this open veranda in an old iron bed
at this moment of time?

The old bath

Two hollows in an old bath
worn by people's bottoms
fascinated Sally as a child.
The old pisé house is stacked full of hay now,
crammed yellow, staring out through glassless windows,
and there's a ghostly army
of thistle sticks, six feet high in the grounds.
It took a lot of lamps flickering
and mothers and fathers and children
sitting in warm water
to wear those hollows.

Impromptu

Every year the weather's unusual.
Stepping from my bed in the dark
I crack the stem of my pipe.
Directing milk jets into a plastic bucket
I'm distracted for a second
and a hoof of polished ebony shifts stance
and tips the white foam on the earth.
That stand of timber all had pipes
(only good
for a collection of giant didgeridoos).
Nothing runs to plan.
Last year who guessed
wool would ride so high?
Could a horseman on the plains of Asia
foresee the numbers of man gone wild?
My father still can't see (he won't listen)
by fencing his large paddock in three
his sheep strip each section in turn
and the feed won't grow back.

We must plan for flaws
(holes to see the sky through
or look at a white horse).
Each generation looks into a new rift,
and history doesn't repeat,
as I amputate the odd tit of a cow with five tits.

The Evening Star

Flickering like a yellow lantern it rises
among an outline of trees on a hill
'the sheep stealer' – Venus, the Evening Star.
('Sheep stealer' to old-timers like Uncle Pat –
'Who's that with the lantern on top of the hill?
Who's that?' – a star.)

At dusk my daughter and her husband
packing clothes and a cot into their van.
As they drive away, all that remains
is a dust haze.

This country cannot hold them –
except with silence.
City,
the son and daughter stealer.
A horse lies on its side in a paddock,
poured out.
Across populations of grass,
with black and green wings
a grasshopper flitters.
This country is haunted,
its children stolen by that Evening Star.

Witnesses

Jehovah's Witnesses, you say with a look of pity.
Well, I'm not one,
but they may well look with pity
at you.

While you and I
despair for the human race
(numbed as reported horrors bombard us)
they can say:
'It is written, this world is mad,
we are not surprised
as we see these ghosts
chasing each other with cutlasses across the quicksands,
the nations sinking.'

They refuse to take oaths –
no man can tell the truth.
Bellboys, plumbers, bus conductors,
they live simply in a time of madness,
accumulate few assets
and wait for the day of judgment.

So many of us die trying to right the world,
the widow in her apartment
shrill with anger at students,
the old socialist cooking toadstool soup for 'The Bosses'.

These Witnesses for all their crazy door-knocking
and Old Testament readings
proclaim one truth –
we are witnesses of the conflagration,
the fires are happening already, all around us.
Our possessions and protests are useless,
our despair is useless.

I am walking down from my father's hill
in another direction
among clean tussocks and granite,
free in the Antarctic night.

Matt Manion

Matt Manion –
matt, dull, unreflecting,
with his small dull fire
cooking chops in the bush –
one day in the kitchen said to my Aunt Margaret
before her spine had begun to curve over
(though her hand since childhood was minus two fingers
laid as a dare on a chopping block,
as her sister Bridge swung an axe),
'You know, Miss McInerney,
I had a funny dream last night.'
'Did you, Mr Manion?'
'I dreamed that you and me was married.'
'Indeed, Mr Manion, I'm very glad it was only a dream.'
'Well, so h'am I. So h'am I.'

The significance of Matt's dream (if any)
hovered unexplained, cut off.
My father called this Margaret's only proposal.

Matt had dinner once with my mother and father,
terrified to find himself eating inside a house,
holding his cutlery like the reins of a dangerous horse.
When she found no peach stones in his plate
Mother, surprised and concerned, asked,
'Didn't I give you any peaches, Mr Manion?'

Matt would have cut a dash
when mosses and ferns ruled the earth,
cooking his chops beside a mild Devonian sea
which was only beginning to taste of salt.

He is one of our guests.

Heat

114 in the shade.

Heat that eats at the very soul –
the early Fathers were right
locating Hell in a hot place.

Plants suffer,
fray in the hot wind and cease to exist,
or else retreat into themselves
in the long siege of the heat.

It becomes a struggle against death
watering the animals twice daily,
dust rising from my boots
as I empty water into the drinking troughs.
The animals suck it straight up,
and I watch for scrub fires.

A neighbour with a pretty complexion
now has a face as florid as butchered meat.
The women are too exhausted to cook or clean,
sweating in armchairs in darkened houses.

I feel sorry for babies and small children.
Some die.

They have no mental resources to fight the heat.
The heat to them must seem a permanent condition,
the world a place of continuous fire.

I remember Sally with beetroot face as a baby
calling out in the hot dark,
staggering around her cot
dazed, like a trapped animal,
wanting to get up and play at two in the morning,
folding up in odd corners and panting.

It's night – heat without light,
insects droning and shrilling deliriously.
At two I get up,
boil some tea, answer heat with heat.
Drinking scalding darkness from a cup
concentrates my mood,
gives form to the void.
Lying back again on the veranda
I wait for daybreak,
my mind holding a small reserve of water,
shrinking drop by drop.

Flowers in the heat

Our native plants know our summer is cruel.
They flower in mid winter
and early spring.
To withstand the heat
unrelated families
adopt similar uniforms,
myall and ironbark (legume and myrtle)
march in the same blue-grey regiment.

The odd mavericks flower in midsummer,
Christmas bells with their rubbery flowers
of red painted yellow,
and some gum trees –
with their massive control of the landscape,
the big tap roots fetching up underground water,
they choose to give nectar when days are fierce.

A letter from the place of pines

I was born at a place of pines
not far from a place of stones.
There's a town built at the place of stones.
That's where I meet people and go to weddings
and buy and sell,
but the place of pines is my permanent address.

At the place of stones there's a red brick church,
a bridge and willows by the river.
At the place of pines there are rusting cans
and fowls sitting in the dust
and a wagtail that sang all last night in my poplars.

At the place of stones there's a feed mill
and a broadcasting station.
They worry about neighbouring towns developing.
At the place of pines some of us go mad.
Ted Hutt who grew the fabulous tomato plant
shot his brains out in a tree.
My slow neighbour Nat, stickybeaking
was told by the policeman to scoop them all up.
But there's not much development.

In the place of stones the houses stand in fenced allotments,
there's a high school and a golf course,
and a mad woman tidying up scraps in the street,
screaming obscenities.

People in both places are much the same,
live under the same moon.

In the place of pines
my neighbours' properties are blowing away in the sky,
and there's a lot of dust flying past
I can't identify
from places hundreds of miles further west
(also overstocked).
This dust blows into the place of stones.

In the place of pines
there are damp patches on linoleum
where my dog Tom has licked up food-scraps,
and there are dead branches lying around
they'd collect for firewood in the place of stones.

The place of stones and place of pines
are both part of my mind.
Travelling between them
I stay sane.

Water from my face

Each year spring occurs
with such vehemence
it's clear the plants don't remember
last summer's dried sticks.

The young discover their bodies
as no one else ever did.
The old say 'We remember'
and forget.

Nature has no memory
and asks no questions.

The water falling from my face
into a rusted enamel dish
(washing under a kurrajong tree)
doesn't recall its shape for long.

Iron and calcium

At night a dozen white cat's eyes
the size of plates,
stare at me down the slope
as I drive up to our house,
the cars and trucks of a lifetime
becoming tons of rust.
(The duco bubbles in lichen patterns
and wild oats grow from cylinder heads.)
When sheep are nibbling grass over our hearthstone
and Spring Forest
is a fine layer of carbon in clay,
there'll be no iron deficiency
in the soil around here.

My neighbour in the pub
said to some bearded scientists,
'So you're up here investigating kangaroos.
Don't waste your time.
How's the joey born?
It's born on the tit!
I've seen it
when it was just two eyes on the end of the tit.'

My neighbour's biology
was set, like his drinking habits.

There'll be no calcium deficiency
in the graveyards of our Central West.

Alpine herbfields

I don't care for alpine landscapes in winter –
I'm more interested in the dead grey leaves
from last summer
pressed flat under the snow.

In summer the herbfield is a coral garden
of constant minute activity
reticulating among buttercup mats,
and portulaca stems, a mesh of red worms.

The granular soil glistens
with moisture,
and the leaf clumps of snow daisies
are a patchwork of silver
up the brown and green slopes.

It's a run-off for dwindling snow banks,
basins of tan rock
spilling into pebbly creeks
through fenlands of cord rushes
where yabbies scavenge in the mud and gravel
of raised ponds.

There's nectar in the shallow tubes
of scented star flowers
and the small brittle flowers of heaths

for bogong moths when they swarm
and leave their skeletons in the grass.

Blizzards and snow avalanches
have shaped the botany.
Snow gum trunks
splay from a root
poured like white and brown lava over rock.
Herbaceous plants won't trust
a single stalk that ice may snap.
Their stems are massed and squash like wire.

If I could, I'd paint snowgrass
on the feldmark pebbled with rocks
like hundreds of sheep asleep.
I'd place a damselfly
among orange flower tufts,
amd eyebrights against speckled granite.

Midsummer snow may bury
a white buttercup overnight.
It's a poor-rich landscape
polished and scrubbed by hardship,
and the terraces of herbfield flower
with gratitude
clinging to their cold mountain gravels.

The meat safe

The day after Jack was killed
a stranger drove
five hundred miles, distraught, to see us,
someone my brother had promised to marry.
We were family she had lost.
She left a suitcase of his starched shirts.

My daughter stayed later at her house,
a large old place,
as the leaves were starting to colour for autumn,
but couldn't sleep for thinking of the birds,
the diamond sparrows, fire-tailed finches
swarming through the old orchard,
three nests to a tree,
and apples stacked loosely in a shed,
and a carcase in a meat safe hanging
beneath an oak
to cool and stiffen.

A child couldn't sleep
for birds and the smell of apples.

Baking at night

You don't get bread these days
with blue and green beetle wings baked into it
and pink stains from some crimson bug.
On hot nights
the lights of the bakehouse drew
all the insects of Waugoola Shire,
and strolling past you could smell the dough.
But they've given up baking at night.

You don't see the fires of the bagmen
under the bridge by the river.
They're extinct too.
Mr Long sometimes humped his swag
for far-off places,
drinking methylated spirits, shadow boxing
and trying to kiss people.
I've tasted his johnny cakes,
flour mixed with salt and water on a fence post
and cooked on a sheet of galvanized iron,
zinc curling off around the dough.
Burned specks turned out to be mouse dung.

After his long tramp across One-Tree-Plain
with a 'cigarette swag'
Jim Long (Old Quizzer) dossed for some weeks
with a dozen other bagmen sprawled drunk
under the bridge at Darlington Point.

He got some meat scraps
and cooked soup for them all in a kerosene tin.
A bagman's three-day-old corpse
when it was noticed
was christened 'Hot and Juicy'.
The bagmen dug a hole by the side of the river,
a bucket of beer
was sent down from the Punt Hotel,
and Constable Brindle read the burial service.

You don't see many drunkards, wanderers
or blind people
(like Mrs Stinson – as children we loved
to see her holding her missal upside down
in church, poor woman).
There's no Cancer Joe for children to taunt.

If I wanted to join the bagmen by the river
under the weeping willows
I'd find no one there,
only the rumble of semi-trailers crossing the bridge,
the big headlights hurtling over.

We live in very moral times.

With the stars as my bed lamp

Olive, I'd like to wake up under a blackberry bush.
I'd like to go on the road.

Pork chops cooked on a shovel –
Mr Long's favourite method –
it ruins the temper of the steel
but I've never tasted anything as good.

There's something about the moon
just rising above a paddock of stubble,
the light on the dried stalks
and among my silver poplars
that bodes ill for the tax papers on my rickety desk.

I'll be one of those men
of whom Mr Long said,
'He's seen more dinner times than dinners.'

Olive, I can see a blackberry bush and a road.
Olive, are you listening?

The Palace Hotel

There are some lusty voices singing
and hands clapping
of fine Aboriginal ladies
(in tune with the jukebox)
as I go past their saloon.

The walls are dirty turquoise,
the floorboards sodden with beer and cigarette butts.
The girls entertain black and white friends,
fall pregnant,
and die of poverty and alcohol.

It's degrading, you say (so do I),
but there's something I like
about the vehemence of their despair,
the way they throw their bodies at life
and don't care.
Black people on a winter night
will sit on boxes and kerosene tins
around a big fire
beneath overcast skies that don't move.

You can tell from the way they sing together
they've more compassion
than most Christian congregations.
Walking past I'm stirred by the voices
of girls in the turquoise saloon,

singing and clapping above the jukebox
with such despair and joy —
something we have lost.

The small bat

Lying on the back of my brother's utility
my head propped among chaff bags
I watch falling stars
flicker down the sky,
like the neurons in my brain
going one by one.

Which of those million lights
melted and fell?
There's no gap to say.

Country and western music is playing
from my son's corrugated iron bedroom.
Near ten each night a small bat
visits his room and flies out again.

Paperbarks

There are new life forms, even as we talk.
The long molecules try, abort and re-try.

Most leaves arrive naked,
but the yulan tree's leaf buds
come in brown capsules,
and the leaves of *ficus elasticus*
have pink paper sheaths which drop off.
Most flowers (unlike leaves)
come wrapped: but not the tulip.

The paperbarks have square leaves and round leaves,
they've scale leaves pressed flat along the stem,
and small pin leaves as fine as rain –
mathematical transformations,
delighting in their own geometry.

In praise of fruit

Fruit is the only food (except milk)
that designs itself to be eaten.
A leaf or cow has no wish
to finish up in your stomach.

But it's my will,
not the will of a slab of beef
putting on weight in a paddock
that counts. Some things
we assimilate by force. (It's not nice,
say cows.)
Nice I am not —and yet
I know fruit has a saving grace,
we need this food without guilt,
these nourishing tons hanging
on twigs.

Get your fruit in bulk,
hastily bottling what you can't eat.
(Overnight the smell of sweetness
ferments,
and your boxes of peaches are snowed
with mould.)

At the markets they've wooden crates as high as your waist
full of pears and stone fruit,
big enough to lie in,

but a rocky bed for your spine —
I'd rather sleep one day
in a pumpkin field in midsummer.

Their day in town

Past the third or fourth gate
on marked public roads (dwindling
to an obscure dirt track)
there's a country of broken-down sheds and stunted children
where political ideas
and bailiffs rarely penetrate.
Only the pension cheque
each fortnight on winged sandals
gets past the dogs with pyorrhoea
snarling from rusty kennels.
Sons of unpainted shacks
loll on verandas and drink beer,
it's always smoke-oh time, and weeds
and spiders are the hardest workers.

From this land of the two-headed calf
come Madge and her husband each half-year,
both well over six foot,
each with identical haircut done by the other,
clipped short up the neck to the occiput.
Marching along the footpath
when they come to the store of her choice,
Madge swings her arm out to the right,
hard across Tom, who brakes dead.

Hand-signalling pedestrians
are a sign of an advanced civilization.
They are arbiters of our democracy –
Madge and Tom.

The Spring Forest

Each year we get further away
from the Spring Forest,
the original text.

'Drinking straws' we say,
sipping a milkshake of imitation vanilla
through a thin plastic tube.
My children in summer
used stubble from paddocks
for sipping crushed strawberry water.
These days you don't find tadpoles
boiled up in the washing.

Each year
there are more gaps in the text,
privet in creekbeds
chokes out she-oak,
weeds blot the lettering.

Each place spoke through its plants
and fauna, until we came.

Planting trees in old age

Auntie Bridge and Uncle Pat –
the doors of certain bedrooms
will always be closed.
We speak by not speaking,
like my daughter's diary
hidden in the hollow of a tree
meant only for the wind to read,
and that's how I leave it.

There are certain mad people
whose madness consists of saying
whatever comes into their mind.

Some things I don't wish to know –
how a fine woman wasted herself
on a simpleton
and grew a garden of plants whose names
he mispronounced or didn't know.

Her roses and Dutchman's Pipe have vanished,
and a lifetime of frustration made tolerable
by not being acknowledged.

'What are you planting trees for at your age?'
I asked my aunt aged eighty.
'*Someone* has to plant them,' she said.

JMJ

Standing on his veranda
my father holds the teapot high
above his head
(like the elevation of the host)
and tips the tea-leaves on a geranium patch.

My father looking out
over a hundred square miles of dusk,
the landscape a missal
of darkened grasslands and flushed hills.

Priests came on horseback,
each with a blue enamel water-bottle
in its calico case,
over the hard and fast plains.

Our nuns coached their boys well in football
(habits flapping among the forwards)
but they trained no scholars.

The happy hour

It's cold, but the cold
won't wake the dead in the ground,
not even whiskey will rouse them,
or the friendly glow of the lights
of the Koorawatha Hotel.
I avoid such friendship, passing
farms where they drink more grain
than they grow.

Driving my tractor home late at night,
standing up to keep warm
in my military greatcoat,
I see a figure
on a horse that has stopped,
swaying in the saddle dead-drunk.
I catch him just as he falls.

I roll up my friend in the coat
and bed him down in the roadside grass,
propping his head on the saddle,
and set out for his household of women
with its blaze of angry lights.

As I walk quickly across the paddocks
already the dogs are barking.

Jack Thompson

The smoke travels with a dead match I fling
in an arc from my veranda.
Under the athel tree in dry moonlight
Jack Thompson's Wolseley is parked,
in good running order except its motor's finished
like Jack who is now underground.

Ethel has given Jack's box of worms away,
the worms for his illegal ghost-haunted fishing trips –
illegal
because he had no fishing licence
in a running war with Fisheries inspectors,
lying low among rushes
blowing out his hurricane lantern,
running out the back door as they knocked on the front
('Inspector Farrell from the Fisheries Department,'
I said to Ethel,
and heard a door bang in the dusk) –
ghost-haunted
because Jack would tell companions on his trips
at a certain point in the river
of this farmer drowned there in a flood,
'a small fellow with a white beard'.
Jack was a small fellow
and at three in the morning a small white-bearded ghost
appeared on the embankment.
But where was Jack?

Jack wore the same white beard,
a pith helmet, white silk suit, horn-rimmed glasses
to call on his friends at the Golden Key café
as a 'Professor from the Health Department'.
After a critical run-down of the kitchen
they still didn't twig it was Jack.

How can I talk about Jack
fading in and out of his illusions
which he half believed in himself?
(like any true artist) –
to define the line of truth
dividing
the innermost skin of the pod of the honesty plant
splitting satin from satin.

Mrs Thompson would speak
an elegy of tears
crumpling her handkerchief in the graveyard,
'Now Jack wouldn't like to see you like this,'
I said, patting her shoulder.

Jack had said
'You know, wouldn't it be nice if Ross called in.'
And was dead hours later.

But Jack evades grief
like a goanna running quickly over hot rock
so his temperature stays constant
mixing shade and sunlight

adjusting perfectly to his audience,
but watch that he doesn't run up you like a tree.

Jack's Wolseley is parked outside.
As I take out the engine
and install the engine from the Morris,
Jack is with me opening a cigar box,
showing Aub Adams
his preserved Japanese finger
collected during the war, his own finger
stuck through a hole in the box,
mottled, discoloured and horribly dead.
Aub, a sensitive bachelor, didn't see the joke.

And his cocksparrow ghost haunts Les the grave-digger,
who spent three days with Jack digging up
some army issue iron piping
that Jack knew about, forgotten from the war,
which Jack was going to hide in the river
and sell off piece by piece.
As I drive across the Lachlan bridge
I hear Jack's river-haunting ghost
propound his 'pre-training school' for horses.
'Ross, some time you could cut us
some stringybark rails for the fence?'
For months he talked of his school,
even showed me the block of land
to be balloted by the council.
'A nice job for my retirement.
That chicken shed'll have to go.
And I've got this lady sculpting two concrete lions

for either side of the gate . . .'
One day Jack arrived with Bill Evans
and I cut them some stringybark rails with my chainsaw.
Weeks later it struck me.
The pre-training school, the concrete lions
had been just an elegant fiction
so Jack could get Bill some stringybark rails.

With me, Jack never tried his Japanese finger
nor his ghost by the river:
I kept him up too late talking.
I thought he was always straight with me.
But even while he was flashing me a wink,
he was leading me into the bulrushes.

A hurricane lantern blows out by the Lachlan.
Jack Thompson ceases to exist.

THE NEW HOUSE

Night thoughts without a nightingale

I've never heard a nightingale,
but I've been kept awake half the night
by a wagtail's intermittent
'sweet pretty creature'
like a solitary thread of water
in a parched landscape.

We stepped from wooden hulls
and the stale air of Europe
into a land of biblical want and plenty.
Old Testament words,
from some country of the soul
'drought' and 'dust', 'flood' and 'plain',
became acrid and palpable.
To the British mind
dust was the decay of the body
and deserts a place for God to exile his prophets.
We woke in real deserts where men died
and dust could choke the sky for days on end.
A first settler
battling through scrub
halted at dusk
beneath the escarpment of a vast tableland
etched with the quicksilver
of living creeks.

We balanced a Meissen teacup
near a trestle loaded with cream cakes,
at a garden party
where a brown snake was a guest.
Its polished body
followed a circuitous route among high-heeled sandals
and waxed riding boots,
observed only by a boy
who froze his cry of panic
until the snake was well clear.

We gave up red hunting coats to go
eeling in wet tiger snake country with George.
With the flood came the eels.
Out after dark with a lantern and pitchfork
we'd fling the thrashing spines
over our backs into a hessian bag.
But one night coming home
George emptied his catch on the ground
and a half-stunned tiger snake wriggled away.
George's eeling days ended there and then.

We laid out baroque exhibitions of produce
at country shows,
geometric patterns of red and green apples,
capitals of trussed golden sheaves,
fanfared by orange tubas and yellow trombones
of giant gourds.
Blazing darkly like jewels
jars of honey and preserves in pyramids
were our night at the opera.

The parochial claimed precedence
for our home entertainments and wild foods.
We enjoyed the rich peculiarity of transplanted lives
as exotic became native.

But we left no tradition.

Roughriding an unregistered Japanese motorbike
to the edge of my land
I switch off the ignition.
It's dusk.
An immense and informative hush of insect noises
proclaims my irrelevance.

The nightingale my ancestors abandoned
mocks the flat square miles they chose.
She mocks their brief, provincial history.
The southern Anglo-Celts
shared some ballads and a way of speaking,
but could not hold this land.

My father
glimpsed a pristine botany
and is scattered minerals.
He cannot recognize
the alien features of his grandson,
or the highway down his escarpment.

The new house

Kev Livio, Chris Parris,
names in strip lettering
on bedroom doors of the new house
(that arrived dwarfing roads
in a cloud of dust) –
The wall of the communal room
is water stained from a fall of snow.
What snow fell on the dreams
of the Carcoar dam builders,
migrant workers, young men
saving a nest egg,
older men on the run from their wives?
Judging from the sticky tape
on their bedroom walls
I guess they dreamed of girls in posters
full-breasted, with no clothes.
A fly door creaks open in the moonlight
as Betty (who's on the Pill)
slips down the hallway in her stockings.

On hot nights in my new house
I'm kept awake
by the dreams of young men building a dam
and the ghost of Betty.

Electricity

It's not hard to choose between
low infant mortality and an art nouveau tile.
There's no nostalgia
about women nursing dying children
on finely carved cedar beds
with embroidered linen.

Through the dusk my wife with quick steps carries
from our old house to the new
my favourite pressure lamp.
Moving across the buff paddocks
its light is softly yellow and archaic,
as today
they've put up powerlines through the trees.

I prefer a world
that's modern, vulgar and well lit.

Man and animal

Man and animal have needs that are the same.
The animal chooses for his camp
the tops of hills.
The rising sun warms the rocks
and he looks out over the plain.

The man, too, needs light and warmth.
Returning to a darkened camp
is his most despondent moment.
His world is unmade and bleak.
Then he coaxes
dead sticks and branches into life.
As the first spark eats into dry leaves
his spirits lift,
and soon the kettle is singing.
He talks to his pots and pans
as though they are alive.
He must, or forget how to speak.

The man and animal are enemies
who do not meet.
They plot and elude each other.
Farmers
punching holes in the night with rifles
are baffled,
and the trapper is called in.

He is paid to have no anger.
He knows and admires his enemy.

Hands sensitive as grass to a footfall
place a twig beneath the plate of the trap.
The twig does not break for smaller creatures,
but it does for the dog.
And steel teeth snap shut.

The trapper is given food and wages by farmers.
but his heart is with a family of dogs
among distant boulders at sunset
who were gone when he raised his rifle.

He arrives in a district,
and dingoes stop howling in the hills at night.
He does not share his expertise.
A few simple secrets
provide him with a living.

Someone

Someone, someone
was calling my name
from the tea-tree paddocks,
kinder than a woman,
a person of air
calling to no one.

The O'Brien brothers

The shaky O'Briens
(one shook his head, the other always nodded)
long ago shook themselves into the ground.
They didn't die
of too many birthdays.

Death came quickly
and doctors slowly on horseback
when 'the diphtheria house'
earned its name.
No one has lived there since.
Now the white Banks rose has been torn away
from the pisé walls,
you see sky through the hallway,
and sky shines through the eyes
of some of our local half- and quarter-wits,
their hair cropped so short, cold air
permafrosts their minds.

Out in the paddocks
marked on maps
there are phantom roads and a township
no one built.

Harry Adams

'Who made the world?'
'Goad.'
'Who is God?'
'Buggered if I know, Michael.'
Harry Adams was the only unploughed land
our Inland Mission could find,
but he was bare hills with poor catchment,
like our hills where, riddled with cancer,
he came back to die.
Who could smell a cup of tea from a mile away?
Harry, who'd slope through the door to the nearest seat,
and glue himself to fly paper laid by my brothers,
or absorb a puddle in the hollow of an Austrian chair,
but he made no sign and was not caught again.
Where did I last talk to Harry?
In the main street: he was waiting
to be taken to the old people's home.
Who was with Harry when he died?
Not our Inland Mission
as he knotted his tie around a low-hanging branch.

Bush kitchens

Here's something about our Lachlan valley
kitchen utensils:
some of the Costello camp ovens
travelled from Hill End
across the continent and back.
They'd cook a four or five gallon stew,
and the lids chipped with use.
Now they're iron pots for tourists
in folk museums.

'Don't use that,'
Harry Adams said as someone
shoved a deal stick in the fire
for a pipe lighter.
'That's my custard stirrer!'

When I could run around
on tin roofs without denting them,
our river
was the mother of improvisation.
We drank the Lachlan boiled
with tea leaves.
She washed our enamel plates,
while our cardigans dried our hands
and blew our noses. (Harry Adams!)

Our only dish-washing machine
was George Grogan's:
at meal end he'd call
the dogs to lick his camp oven out.

And kitchens were a moveable
circle of light by the river.

Killing

My mother canned some chickens
in glass jars,
but no one would touch them,
haunting her kitchen
like stewed lungs.
I've seen Bill Sutherland cry
when killing a sheep,
or a calf was born dead.
Men are moody on killing day.
The women know
and feed them sausages.
But next day it's meat.
'The priestly knife,' some city poets say.
Country people kill because they have to.

Drying out

One of those times Mr Long
blew his cheque,
then dried out by the Lachlan,
he chose a site for his camp fire
at some distance from his fellow bagmen.

Spreading slices of stale bread
with no jam from an empty tin,
he poured phantom tea
from a boiling billy
and gave colour with his pantomime
to the iron rations.

He'd spar with us children.
'I can fight like a hare and run like a guinea pig.
Put up your claws.'

Jim Long and John Manion

Shortage of cash starved
Mr Long into celibacy,
long moon-baked nights
warped by the note of a bird.
Each morning John Manion
removed the kettle from the fire
so Jim Long when he rose
had no hot water,
yet each night Jim cooked their dinner.
When Jim broke his leg
John made some crutches
and sharpened the points
one wet summer . . .
Bad luck they lived in a time
before sex and money
were a basic right.

The Lachlan

His kirk hidden by a wall of pines
our Scottish parson
gets rid of his religion in the garden.

Harangued by a Jehovah's Witness
across the fence
Mac in singlet and shorts
leans his mattock on a tree,
and welcomes more discussion – in the house.

The Witness settles in an armchair,
when Mac reappears in his collar,
robed for theological debate.

The Lachlan is my church.
River oaks and gums form ecclesiastical glades
of sunlight
for the dragonfly
scooping a mosquito.
Kurrawongs are Redemptorists
delivering black and white sermons
from high pulpits.
In late winter they hector
blossoms from dull scrublands,
tiny aromatic paper stars
and yellow pollen balls.
Our smaller plants are opportunists.

They temper faith with caution,
flowering in spring before the heat sets in.
But the faith of our river gums is unlimited.
They postpone their blossom for off seasons –
noblesses oblige –
and are the organ pipes of an instrument
with no beginning or end.

Our river is a church with an open door,
but the parishioners are wary and fussy.
Teal nest in thick ground cover,
and musk duck on limbs leaning over the current.
Small finches patrol
the dense understorey
where an eagle's large wings are clumsy,
and blackfish are invited
by moving light and shade,
irregular rocks and snags,
the idiosyncrasies of a known place.

Our river is fed by the spill-off of unstable parishes,
by moveable creeks
unloved by surveyors
who would channelize every watercourse.
Public servants
jogging through city parks in shorts at lunchtime
agree and plan policy papers:
'Channelization and Farm Economics';
'Channelization and the Controlled Landscape'.
Meanwhile fish vote with their fins in denuded creeks.

You don't find our local farmers
jogging around their paddocks,
giving themselves arthritic knees.
Only some streams that feed our Lachlan
run tirelessly,
but with more enjoyment than bureaucrats,
to express certain laws of gravity and motion,
the pleasure of molecules boiling coldly
around a boulder
cutting through clay slopes
to granite bed rock.
We observe
their freeplay and turbulence,
individually random but a massed force
scouring basins around roots, engineering shortcuts,
disappearing under sand beds.

The spiny anteater is afraid of my love
and buries himself in bush sand,
as my boots halt and watch.
Birds on the bank exchange passwords
as Jack, my dead brother, is rowing
and my adolescent hand
trails in khaki water, a translucent ghost.
Baptisms, marriages, masses for the dead
float past, with sticks and a drowned page
of the *Cowra Guardian*.
Lights burn on the waters at night
from votive candles of stars
and the isolated homestead on a hill,

where an ageing brother and sister
inherited celibacy and a farm.

Our river was mother to a tribe
of 'Murrumbidgee whalers' – swagmen
whose well-stocked rowing boats
drifted through a life of ease.
They saw more reflected sunsets than fence posts,
and odd jobs –
the artfully constructed pile of firewood
that was hollow inside –
passed the time between books.

Our river was a great conveyor of people,
strikebreakers disembarking at midnight
from a paddlewheel steamer,
commercial travellers, itinerants, bales of wool,
but now only the odd party of poets
en route to a festival
travels leisurely waterways by outboard motor,
with their notebooks and binoculars.
We forget our river
was the highway for our invading ancestors,
and we seized these flats by force
from a government of naked old men.

I share a cup of tea with the poets on the bank.

We discuss our river's love of decoration,
garnishing herself with sea tassels and water mats,
painting herself with tree reflections

and glimpses of sky,
and the visiting card she leaves after floods:
the smell of dead frogs and crickets.
Our river is bohemian,
the haunt of lawbreakers and loafers,
watching their reflections in waterholes
and a yabby attach itself to meat on a string,
cautious as a stick insect,
and the middle-aged man who furtively greets you,
puffing up the bank, as you descend,
then find the altar he's built –
male genitals sculpted in sand.

Our river is respectable.
She provides cleared spaces for church picnics
and Mrs Mac teetotal and girlishly light-headed
from 'non-alcoholic' cider vouched for and served by Mac.

I yarn with the poets by their camp fire.
My stubble at the end of the day is grey
and they're pink-shaven, black-bearded and hatless.
But they can share a joke with a man who wears a hat.

The river is our confessor,
absolves and dissolves our chemical mistakes,
and our soils are renewed.
But her health is our health.
The flow off of phosphates
feeds a killing blue-green scum of algae.
We can feed our mother too much poison –
our river is mortal.

Yet this evening drinking tea with poets
the river is immortal.
Their boat is pulled up for an overnight pause
in its exchange with the river –
the chatter of water against waxed planks.
A pressure lamp lights a circle of grass
and the souls of the dead join our congregation.
Insects and frogs are a massed choir.
Cooking sausages provide clouds of bush incense,
and we think of the lovers who are married on these banks
in vestries of grass,
experimenting through gaps in clothing.
Her hair is tangled with seeds.
The girl and boy
are vague as they rise,
hardly aware as they pass
of a red glow in the dark,
an old man of the river,
smoking and watching for golden perch and cod
to fill his traps.

Petty crime is an art
which flourishes along these banks.
Lights nod, nets hang from black corks
and ducks are suspicious.
Our river is a church with no gods.
This orchestra of voices from the ribbon weeds
is rich in its mistrust.

Ancient theft

'You've heard of John Gilbert and Ben Hall,
the bushrangers?
He was on good terms with them.
That big rock on his property,
Gilbert's Lookout, it's called.
Gilbert could peer out for miles.
Now how did a young fellow
from nothing – with nothing –
walk into a property
so well set up, with plenty of stock?'
My father asked about his father,
peering from the rock of old age
into the land before his birth.

If I'm the heir of ancient theft
I sup with silver
my grandfather stole from my grandfather.

Old Testament country

The camping place of wild horses was a sign.
Horses found the 'Wash Pen',
the only permanent water
in this Old Testament country.

Where farms now jostle on the map
my father's father selected
horizon to horizon.
There were granite outcrops
and no fences.
The dark-skinned Canaanites
were soon hunted off,
leaving the wilderness
of leaves to make his own.

In Ireland he called himself farmer,
here he was 'on the land'.

Bush wildflowers
were a Joseph's coat of colours,
and wild horses, muzzles to the ground
were a sign

in a time before Jesus
and the products of the city.

Gladstone Watts and the crystal set

The weirdest night of Gladstone Watts
began near sunset on the plains –
arriving at a tent
while droving a mob
with a 'half-Pomeranian' –
that's shorthand for a dog
that couldn't work a chook.
A boundary rider came out,
pumping Gladdie's hand,
talking thirteen to the dozen
(Gladdie wasn't so keen –
he'd spoken to someone a day ago),
and started chopping posts and rails
from mulga; in half an hour
had built a yard for Gladdie's sheep.
After their camp fire meal
the man produced 'this wireless thing.
It could darn well near talk.'

Given the only bed in the tent
beside some books –
where were those voices?
Gladdie kept eyeing the axe by the fire,
and when the man was snoring
slipped out to bunk with his sheep.

There are questions the western plains can't answer,
voices talking from a man's hand,
lights where no lights should be.

'I've figured since,' Gladdie reflects,
'he was a very clever man.'

The two wethers

'The two wethers'
of Violet Hill
grew old and never read
Professor Fowler's 'Science of Life,
Including Love,
Its Law and Power'.

The basalt tablelands produced
no pair of sisters
eligible for their sheep run.
Farm economics required
a double match, or nothing.

On Sunday nights they joined
in hymns from 'Scottish Paraphrases',
sung only to a tuning fork;
and rode home
through mist and frost.

Nothing of two old brothers
survives, no anecdote:
only a name.

Poverty bush

We pledged our vows by the poverty bush.
I prepared for your hair
a garland of wiry stems, blue leaves and thorns
that are the first to appear
on scalded land.
At the technical college in town
they're learning the poverty bush dance,
one step forward and two steps back.
When sixty-year-old men ran short
my friend Heather was the first to revolt.
She walked across the floor to Wendy
without self-consciousness,
and watching the movements of the males
guided her among the couples.
Some girls won't get up without a man,
so they sit not budging all night.
But most enjoy dancing with their sex
when gents are scarce –
both step forward, no one steps back.

Law

Law is History, not Science.
We think analogue, not digital.
The same thought travels many dendrites at once.
The brain is a consensus,
not a hierarchy.

Because my son-in-law
wrote a letter
citing a case
about furniture bailed during London air raids
and unclaimed for years,
my antique steamroller
left by me for a decade in someone's yard
came back from its innocent purchaser.

The racket of its arrival
woke the landscape
like a twenty-one gun salute,
and grandchildren ran out to greet
the vast prime mover,
two storeys of red and chrome metal
pulling along our dirt road
through a frame of willow trees,
a bridal train,
with the sun-dappled bride, my steamroller,
chained and suspended
on the long low loader.

The women and children withdrew
when the business of unloading began,
of easing the steamroller's treacherous gravity
from the low platform onto the earth.

Burly and red like their prime mover
the men asked quietly,
'Any women about?'
Then unzipped
to expose hidden parts
that drizzled and hissed steam
in dust and dead grass.

Modesty is unwritten law
as precise as legislation.
It constipates women in Bihar by day
who fertilize their fields
only after dark.
It decrees seclusion
for the baboon giving birth.
Only the father can approach.

The law has brought back my chattel,
but she is wearing
an unfamilar brass horse 'Invicta' –
and one morning it's gone.

Nature, like law,
builds with precedent.
The foetus
discarding its gills and tail

has to relive evolution
to become human.
My steamroller and I
are deposited here
by rules that are not simple.

The golden wall

Don't ask Uncle Pat why the night sky is dark –
in hot weather
taking his mattress out on the grass
inside his dog-proof fence to sleep.
When Pat lifts his face up to the night –
propped on a pillow
of kapok stuffed in mattress ticking –
he'd fix you with sheep drench if you told him
that his line of sight
should intersect at every point
with a near or distant star
glimmering in the transparency of space
so the whole sky
should be ablaze from end to end
like 'a golden wall'.
Pat's golden wall was his orange tree.
Like Uncle Pat it had never borne fruit
until I dumped five tons of chicken manure
on its roots.
His line of sight
from the cane lounge where he sprawled
intersected at every point with oranges
twenty feet up in the sky,
a Utopia of fruit
which the district came to visit and eat,
oranges with no ending
like the return veranda

around the four sides of his house
where nephews and nieces ran forever
and their children after them.

Pat forgot his promise to pay for the manure
and the oranges didn't come back.
But he didn't miss them,
so don't ask Pat why the night sky is dark.

Olbers' riddle has hung around
for centuries.
You can't explain it by absorption.
Gas and dust heat up and glow.
Nor by absences or voids.
Every square inch has its galaxies.

Ask the cells inside your head
the same riddle,
why don't they all blaze at once
a golden wall of noise,
each neuron singing its own note
deafening your mind with light.
Political and religious visionaries
promise us this,
every cell singing in unison,
a mass of indistinguishable stars.

But something in the universe denies
the golden wall,
some structure which became Uncle Pat
calling to his nephews from his cane lounge,

'Now don't trample them tomahawk plants!'
(meaning hollyhock plants).

Pat prefers his own company on hot nights
leaving Auntie Bridge inside
with pictures of saints on the bedroom wall.
He takes his bedding
and lies in a darkness
where each star can broadcast as a soloist.

The universe
is a composition of unique bodies
on display,
and the night sky of the mind
allows a single file of thoughts
to light up as a sentence.

Prickly moses

The weather comes from somewhere else.
The rains forgot to come.
It's like our lives, we say,
bitterly knocking tobacco ash
from a pipe on the stone hearth.

Driving late at night
past abandoned homesteads
(a rusting sign 'Byzantium'
hangs from a gate)
there's not much for the banks to sell,
some thousands of desiccated acres
and thorny wattles with yellow pollen balls,
the gift of thin twigs.
Our diaries can find no pattern.
No one stole our rain.
I stop and get out
in the empty stillness
of the last continent
where men grew plants.
Our weather starts here
in these plains and hills that we chose
for their irregular rhythm.

Gilbert and Hall

On the road between Boorowa and Binalong
they are inventing morality.
Holding up a black-bordered envelope,
John Gilbert tells his captives, 'We must respect death,'
and leaves it unopened.
He fancies a slice of wedding cake
in a letter.
'Don't,' says Hall. 'It might be poisoned.'

After they've stripped all the banknotes
from the mailbags
Hall is for burning the letters and cheques.
'No,' says Gilbert. 'They might be useful
to the owners.'
So they leave them scattered in the grass
for the mailman and a passenger to collect.

O'Meally, the mate of Gilbert and Hall,
has his morality too,
from his saddle pumping shot
into Barnes, the shopkeeper,
who is galloping away unarmed.
'I'm sorry he's dead,' O'Meally says later.
'But it's his own fault. He should have stood.'

Gilbert and Hall are actors in a self-scripted play.
Drinking port in the firelight with victims

after he's robbed their house
Ben Hall bounces their child on his knee.
'I'd love to take this young 'un on the track –
and make him a man.'
'No, no, you can't, you wicked fellow,'
the mother says.
Happy Jack Gilbert commands
the grown-up daughter
to sit at the pianoforte,
and predators and prey sing together.

They are specialists who love their work,
on stolen racehorses
outdistancing police
whom they outgun
with five or six repeaters
slung in holsters around the midriff.
They laugh and discuss the habits of victims,
the miners who challenge them to 'a fair fight –
show us your fists',
the squatter who indignantly rejects
the silver change they give back,
the merchant who sneaks a sovereign into his boot.

They have rehearsed the last scene.
One of their 'harbourers' for a reward
will lead a group of shadows
to a lonely hiding place
chosen for its dense scrub
where a fugitive can vanish like water.

At dawn a posse of troopers will open fire
on a man scrambling into his boots.

They relish bartering at gunpoint,
a hat, a cloak or saddle.
Death is three days away from Gilbert.
'That's my horse you're riding,'
Happy Jack jokes with his victim.
'You're a common horse thief!
I've a mind to lay a charge.
But I'll swap you this,
if you give me back mine.'

They delight in arbitrariness,
stripping an old wayfarer of his life's savings,
robbing a child's piggybank,
staging a party at gunpoint
in Robinson's Hotel.
There are carts and carriages lined up outside –
with valuables untouched.
The captives drink hard liquor
shouted by Gilbert and Hall
who abstemiously sip at bottled ale.
Happy Jack signs
three written passes of leave for an hour.
On the hour Hall goes after them.
'It's not right,' he says
as he finds the men strolling back.
'We gave you one hour's leave
and you're ten minutes late.'

Driving the Boorowa road a century later
the morality of Gilbert and Hall
is the false apple scent
of roadside briars,
inspiring middle class historians
to romanticize common theft
and paying off a network of spies
as robbing the rich to give to the poor.

I pause to drink tea
from a stainless steel thermos
and the hard good looks of Hall
glitter from translucent thorns.
Happy Jack is cheeky as the pink and white flowers
of rosa canina.
'You've stolen my life,' he says.
'They're not your wife and children. They're mine.
Coming home tonight
you'll find me by the fire
sitting in your sheepskin-covered car seat.
I'll be tapping your rosewood pipe on the hearth
and your dog will be sleeping at my feet.'

Jack again

Growing older
the prime numbers come less often.
About to tell you something
at night my thoughts insist
on travelling branch lines
they closed down years ago.
Summer weeds sprout from the bluemetal track.
The rusting rails are two scorch marks
that converge in shimmering heat.

It's fifty years since you dragged yourself,
leg gouged by an axe, felling trees to this spot.
A medical student, not trusting country doctors,
you didn't come home
and flagged down a train for Sydney.

Not yet found

I chose the name Spring Forest
and I've yet to find the spring.

Some unfinished equations
are the closest I've come
to the puzzle of why I'm here.

There is a book before our eyes –
the night sky of the universe.
Galileo saw its language was mathematics.
A cricket's encrypted love song,
light from an ancient star
are mathematical messages
arriving in sultry air.

Imaginary and complex numbers
allow life to reproduce itself
endlessly and intricately
without repetition –
the elusive algorithms of a summer night.

Invicta

She is Invicta – head of the horse,
long eyelashes and brown melting eyes,
currency from the copper age.
We are lined up, straggling youths on ponies,
a parody of cavalry at the start of a campaign.
We are lifted through the air,
she has taken our hearts.
Six thousand years ago we mounted her,
plains-dwelling pedestrians,
and she taught us theft and war.

The two tents

Late and pragmatic
we arrive with Mr Long in his story
past dusk at two tents on the plain
(a mother and daughter
up from Dubbo on miners' pay day).
Talking to the mother in the first tent;
he finds she's had no customers
and the 'titter' six or seven.
He tells the mother,'You'll do me.'

Two tents are camped on the plains,
Hygiene and Romance,
and the flap of Hygiene closes,
welcoming Mr Long.

Botany lessons from Mr Long

Gifts of the mind, not facts, survive.
We would stand under the trees-of-heaven
on the dry creek bed of cool grey pebbles
(large as rocks with not much water to wear them down).
'I think I need a new pipe,'
Mr Long would say.
'I'll go on an expedition
to that pipe tree.
A few months back
it had some pretty good pipes ripening on it.
They'd be just about ready by now.'

'Mr Long,' we'd all chorus.
'Will you pick us a dinky
from the dinky tree!'
A tree hanging with tricycles grew
in a place known only to Mr Long –
and one day they'd be ours.

'Well, I won't be going near there,'
he'd say. 'And last time I did,
the dinkies were all very green.
It takes a long time for dinkies to ripen.'

On a grey morning the bush
would swallow Mr Long in his moleskins,
as he set out for his pipe tree.

The end of the Pacific War

We were back from a war.
POWs, tall and emaciated
walked unsteadily from ships.
It was spring.
Women were scrubbing brown air raid paper
from window panes,
the neon signs of shops were switched on.

The end of my Pacific War
was rowing out with Jack on Sydney Harbour,
bobbing in a purloined dinghy
among ships and fireworks,
the acrid smoke of detonations
harmless now.
The black and glittering jelly of the harbour
blazed with exploding chrysanthemums
and blue and green starbursts –
an innocent parody.

That spring night in the dinghy
we watched the theatre of the mind,
a spark firing
a consensus of neurons,
the oscillating bursts
of co-operative energy
igniting and fading in the sky.

Later among partying crowds
we wandered 'the Cross',
excited by the turbulence
of animal and flower scents
and chasing
some tremendous discovery.

Bell tolling, a fire engine drove past.
Jack hailed a taxi.
'Driver, to the fire.'

A party of star gazers

I doubt I'll be the first to report
a stranger lighting the sky,
a moth's eye flaming,
the birth of a supernova.

As the shockwave of neutrinos registers
in water tanks miles underground,
my brass telescope sits on top of a cupboard.

I'm not a disciplined amateur,
watching Jupiter night after night
for subtle colour changes,
interpreting small differences.
My brass telescope
is a family occasion
for husbands and wives and children.
With a heavy duty electric torch
trained on the ground
and deliberately thumping feet
to warn my friends, the venomous snakes,
we walk out into the shockwave of cold air,
under an expanding or a steady sky.
We erect the telescope with its tripod
on a gravel crest
looking across the new garden's dense tea-trees
regular as topiary, housing by day
a parliament of honey-eaters and parrots.

Children jostle to look through the eyepiece.
Adults stand back, interested.
I point out obvious features –
the rings of Saturn
that will granulate
after months of patient watching –
and craters of the moon.
Fine details,
a lava flow across the rim
of an impact crater
resolve themselves
for the dedicated observer.

But not for this audience
of talkative appreciators,
sceptical but not technical.
Our breaths are suspended
like the icy detritus of comets.
On the first night of autumn we stand
some in cotton, some wearing wool,
enjoying the riddle of the sky,
not comprehending theory
that says everything came from nothing.
We know our fate is enacted
in light refracted through these lenses.
Blinking, our eyelashes brush
our beginning and our end –
or endless continuity.

The Daisy Picker

Send my corpse home on 'the Daisy Picker'
and bury me in my pyjamas –
per 'the Daisy Picker'
because it's so hated and loved
for its procrastinations.
Passengers alight
and pick wildflowers by the railway line.
Then with a shuffling of buffers
and whingeing of couplings
it startles away with no cause.

The shadow of our 'Daisy Picker'
crossing a bridge
intersects the sun,
as I float on my back,
ears tingling with pressure in the Lachlan.
Cow's toenails and bones would sink.
It's a matter of displacement.
We're judged by quantities.

But don't give my measurements yet
to our local undertaker –
carpenter's rule in his pocket,
as he sells me canaries.

I'm going more trips
on 'the Daisy Picker',

journeys with an end
but no destination,
as a red dragonfly
paces the train.
You'll see my face lean from a window,
shaded by a hat brim from the sun,
observing rocks and summer weeds
advance as they recede.